ultimate labs

helping you save lives

Your CliffsNotes™
for USP Water Testing

Simple Techniques to Avoid Facility Shut-Down

5940 Pacific Mesa Court, Suite 209 San Diego, CA 92121
www.ultimatelabsinc.com| info@ultimatelabsinc.com | 858-677-9297

Why A Guide to Compliant USP Water Testing?

We'd like to talk candidly with you, from one technical operation to another. You know you have a great product. Your company owners and team members all have big aspirations for your business - your product will change and/or save lives. AND, your purified water is the lifeblood of your product(s)!

That last statement about the importance of your purified water may or may not ring true to you, but it should. Chances are you are spending way more time and energy than you need to in order to remain USP purified water compliant.

This guide will help you better understand the world of purified water (used in facilities such as yours), and it will give you some insight on how more effectively to manage your compliant water testing process.

THIS GUIDE WILL HELP YOU BETTER
UNDERSTAND
THE WORLD OF PURIFIED WATER

Why Do We Do the Things We Do?

We, at Ultimate Labs, save lives through testing. With over 25 years of experience, we apply our knowledge of microbiology industries that impact consumer health. These include environmental monitoring and microbial, water, and food testing for the pharmaceutical, biotech, medical device and food industries.

Over the past year, we have tested dozens of products, hundreds of water systems, thousands of clean room samples and as a result, saved over a million lives. We have created process solutions that ensure samples custody, efficient testing and professional and personal communication. When working with us, you will encounter a complete client experience dedicated to your success and your mission to save lives..

WITH OVER
25 YEARS OF EXPERIENCE

What Have We Created?

When you work with Ultimate Labs, you can leverage our investment in the latest technology and personnel. Our financial commitment to staffing and equipment maintenance and improvements ensures accuracy, quality and efficiency.

Here are some improvements we've made from listening to industry needs and client requests:

- BiomerieuxVitek® MS – Developing mass spectrophotometry-based tests for rapid and accurate microbial identification
- Ulti-Pak – Proprietary custom plate holder and carrier system to ensure chain-of-custody and sample integrity
- Project Management Program – Your designated point-of-contact partner that tracks all aspects of your test process – from sample collection to interpreting results and reporting

Solutions Tailored for You

We become part of your team!

Consider us your outsourced in-house laboratory support. Like you, we move along with advances in the field, expanding our capacity to solve problems, and be an asset for our clients.

Our routine tests include environmental monitoring, water testing, food testing and microbiology testing.

THE CREATIVE & EFFECTIVE USE OF PLAIN ENGLISH

All too often in our industry and the industries we serve, language becomes too technical and complicated. Wouldn't it be nice to have something explained to you in **Plain English**?

Well that's what we're gonna do in this "very technical" guide!

Effective and compliant water testing can be very technical, and in our opinion, over complicated. In the pages that follow, we are going to take you into the bowels of the Pharmacopeia and explain exactly what you need to do, how to do it, and what it means for your operation. And we're going to do it in **Plain English**!

spend so much time, money, and effort on water testing.

Oh, and by the way, if that is indeed the case for you. Don't worry you're certainly not the first (or only) one that feels that way.

Okay, here we go, **The Ultimate Guide to Compliant Water Testing in Plain English.**

Your Operation's Purified Water System – Where to Start

So what's the big deal about purified water? Whether you know it or not, water is the life blood of your entire operation. But we're sure you knew that already because so many people up and down your company's organizational chart are concerned about your purified water, and we're willing to bet that includes you too! Are we right?

For any biotech product manufactured in the United States (and this includes a wide range of products), the water used must meet certain standards created and determined by the United States Pharmacopeia (USP).

Oops! We just had to use one of those big (non-Plain English) words. So let's quickly explain what a **pharmacopeia** is. A pharmacopeia is just fancy word for a reference book for Pharmaceutical products. It's a book (or manual) that's updated annually by a nonprofit organization. In this instance, the USP we're referring to is created and updated by a non-biased group made up of doctors, scientists, and other medical professionals in the U.S. Together, they determine what the standards of biotech products, and their component ingredients, should be.

By the way, USP has no role in enforcing its standards; enforcement is the responsibility of the FDA and other government authorities in the U.S. and elsewhere.

The USP is broken down into many different formulations, instruction and methods. Each formulation lists the main ingredient, the characteristics it must have to be compliant, and it even spells out how the ingredient must be tested in order to determine if it meets its standard.

YOUR "RECIPE" FOR SUCCESS STARTS HERE!

Each one of these formulations is called a **"monograph"**. A monograph is really just fancy way of saying this is how you make it and how you know you made it properly. It's kind of like a recipe.

So let's crack open the USP and drive right in to purified water testing and compliance!

When you open the USP to the monograph for Purified Water, this is what you see…

Purified Water

[NOTE—For microbiological guidance, see general information chapter *Water for Pharmaceutical Purposes* ⟨1231⟩.]

H_2O 18.02

DEFINITION

Purified Water is water obtained by a suitable process. It is prepared from water complying with the U. S. Environmental Protection Agency National Primary Drinking Water Regulations or with the drinking water regulations of the European Union or of Japan, or with the World Health Organization's Guidelines for Drinking Water Quality. It contains no added substance.

[NOTE—Purified Water whether it is available in bulk or packaged forms, is intended for use as an ingredient of official preparations and in tests and assays unless otherwise specified (see *8.230. Water* under *8. Terms and Definitions* in the *General Notices and Requirements*). Where used for sterile dosage forms, other than for parenteral administration, process the article to meet the requirements under *Sterility Tests* ⟨71⟩, or first render the Purified Water sterile and thereafter protect it from microbial contamination. Do not use Purified Water in preparations intended for parenteral administration. For such purposes use Water for Injection, Bacteriostatic Water for Injection, or Sterile Water for Injection. In addition to the *Specific Tests*, Purified Water that is packaged for commercial use elsewhere meets the additional requirements for *Packaging and Storage* and *Labeling* as indicated under *Additional Requirements*.]

SPECIFIC TESTS

[NOTE—Required for bulk and packaged forms of *Purified Water*]

- **TOTAL ORGANIC CARBON** ⟨643⟩: Meets the requirements
- **WATER CONDUCTIVITY**, *Bulk Water* ⟨645⟩: Meets the requirements

ADDITIONAL REQUIREMENTS

[NOTE—Required for packaged forms of Purified Water]

- **PACKAGING AND STORAGE:** Where packaged, preserve in unreactive storage containers that are designed to prevent microbial entry.
- **LABELING:** Where packaged, label it to indicate the method of preparation and that it is not intended for parenteral administration.
- **USP REFERENCE STANDARDS** ⟨11⟩
  USP 1,4-Benzoquinone RS
  USP Sucrose RS

USP Monographs

AND THIS IS WHAT YOU'RE PROBABLY THINKING!...

Purified Water

[NOTE—For microbiological guidance, see general information chapter *Water for Pharmaceutical Purposes (1231)*.]

H_2O 18.02

What!? Another Standard? I can barely translate this one and it's supposed to be easy!

DEFINITION

Purified Water is water obtained by a suitable process. It is prepared from water complying with the U. S. Environmental Protection Agency National Primary Drinking Water Regulations or with the drinking water regulations of the European Union or of Japan, or with the World Health Organization's Guidelines for Drinking Water Quality. It contains no added substance.

[NOTE—Purified Water whether it is available in bulk or packaged forms, is intended for use as an ingredient of official preparations and in tests and assays unless otherwise specified (see *8.230. Water under 8. Terms and Definitions* in the *General Notices and Requirements*). Where used for sterile dosage forms, other than for parenteral administration, process the article to meet the requirements under *Sterility Tests (71)*, or first render the Purified Water sterile and thereafter protect it from microbial contamination. Do not use Purified Water in preparations intended for parenteral administration. For such purposes use Water for Injection, Bacteriostatic Water for Injection, or Sterile Water for Injection. In addition to the *Specific Tests*, Purified Water that is packaged for commercial use elsewhere meets the additional requirements for *Packaging and Storage* and *Labeling* as indicated under *Additional Requirements.*]

SPECIFIC TESTS

[NOTE—Required for bulk and packaged forms of Purified Water]
- **TOTAL ORGANIC CARBON (643):** Meets the requirements
- **WATER CONDUCTIVITY,** *Bulk Water* (645): Meets the requirements

ADDITIONAL REQUIREMENTS

[NOTE—Required for packaged forms of Purified Water]
- **PACKAGING AND STORAGE:** Where packaged, preserve in unreactive storage containers that are designed to prevent microbial entry.
- **LABELING:** Where packaged, label it to indicate the method of preparation and that it is not intended for parenteral administration.
- **USP REFERENCE STANDARDS (11)**
  USP 1,4-Benzoquinone RS
  USP Sucrose RS

USP Monographs

Huh?!

All of this means, if your water comes in contact with your product or anything used to make your product, *it is regulated!*

Yes, The USP is a master at cross-referencing! This NOTE references microbial (bioburden) testing. We'll talk more about this later.

Where are these?!

More?!

Make sure you package or store your water in clean or sterile containers.

Label your containers as **USP Purified Water.**

Reference standards are for the chemical tests <643> and <645>.

Another masterful cross-reference to the chemical testing section of the USP. Not only are there monographs, but also numbered tests!

Plain English, Please…

In a nutshell what this particular monograph is telling you is that three distinct and separate tests need to be performed on your water to verify that it meets the standards for purified water. Two of the tests, Total Organic Carbon and Water Conductivity are stated right in the middle of the page. But I'll bet you can't figure out what the third test should be!

Find the Hidden Test in Your Standard

The USP likes to "hide" things within other things. If you read the Definition section of the monograph you will get your first clue as to the 3rd test. The definition references microbial contamination which slyly refers to Bioburden testing - the 3rd test.

Because purified water is the lifeblood of your facility and, therefore; your products, and since we now know it takes three different tests to verify the compliance of your purified water against USP standards, we'd like to give you an easy to understand explanation of:

(1) what each test measures,

(2) how the test is performed, and

(3) some of the pitfalls in performing these tests.

Prevent Your Facility from Being Shut Down – Discover Pitfalls

Knowing what the testing pitfalls are, and how to avoid them are a key to successful, compliant water testing. Just think of it this way - what would the cost be of having to shut your facility down for a day or two? - Okay, now I bet I've got your attention again! Let's continue…

Conductivity

Conductivity is a measure of the concentration of ions. It also measures the water's capability to pass electrical flow. The greater the concentration of ions the greater the capacity to pass electric flow. Conductive ions come from dissolved salts and inorganic materials such as alkalis, chlorides, sulfides and carbonate compounds. As an example, and a point of reference, sea water is approximately 50,000 times more conductive than distilled water.

Type	Electrical Conductivity (μS/cm)
Pure Water	0.05
Distilled Water	1
Rain or Snow	2 - 100
Surface / Ground Water	50 - 50,000
Seawater	50,000

When testing for conductivity in Purified Water a calibrated conductivity meter is used. The USP lists 3 sequential tests or stages. If the water passes the first test no further tests are required. If, however, the first test fails, the water must be subjected to the 2nd test. If the sample fails the 2nd test, the third test must be performed.
If the water fails, the 3rd test the sample fails.

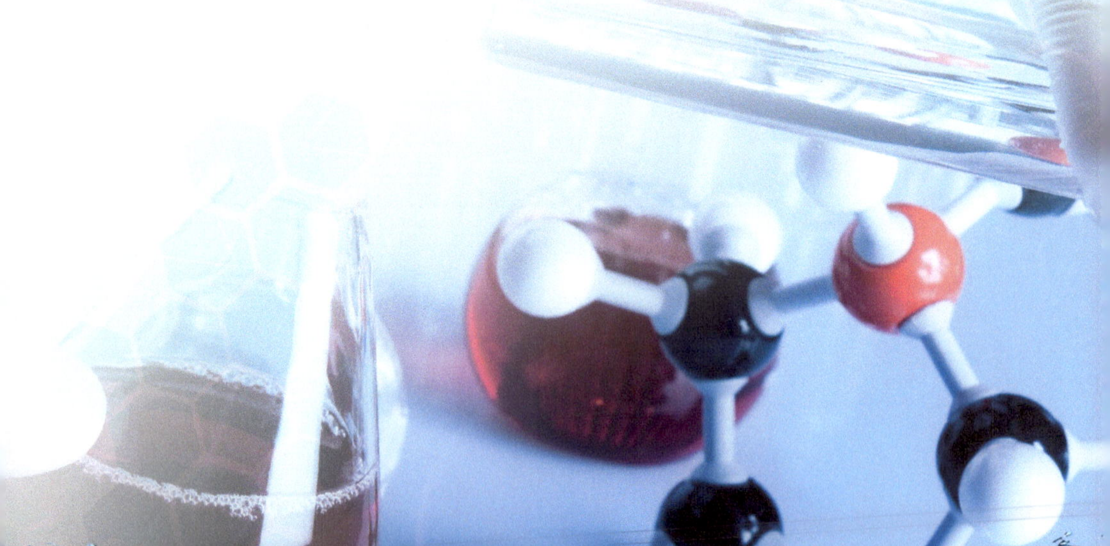

Avoid Loss of Manufacturing Time and Dollars!

Some of the pitfalls and common reasons for failing the conductivity test can be attributed to equipment. Common causes of water conductivity failure are:

- dirty or worn polishers
- worn out membranes
- air contamination in the water line
- over-used resin beds

Implementing an effective preventative maintenance can reduce problems in this area.

Total Organic Carbon (TOC)

Total Organic Carbon (TOC) is a term used to describe the measurement of organic (carbon based) contaminants in a water system. Organic contamination can come from a variety of sources, since "organics" are compounds such as sugar, sucrose, alcohol, petroleum, PVC cement, plastic based derivatives, etc.

- Organics may exist in the feed water.
- Organics may result from the leaching or shedding of various components within the purificat ion or water distribution system.
- Organics may result from the formation of biofilms (bacteria) in the water system.

Total organic carbon (TOC) analyzers can determine the TOC level by measuring the leftover carbon (organic) after removing the inorganic carbon, or by measuring the difference between the total carbon and the total inorganic carbon to determine the actual TOC level.

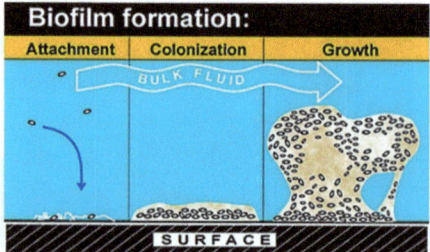

Save Time and Money with This Important Tip!

The pitfalls to avoid when performing a TOC test are primarily related to the cleaning solutions used for cleaning the system. Many cleaners and disinfectants are comprised of organic compounds. For example, a solution such as Isopropyl Alcohol (which contains organic compounds) is often used for disinfecting gloves. The same gloves used in obtaining the sample. Another cause of high TOC readings is a presence of biofilm in the water system. This requires a system sanitization to remove the hazardous material.

Bioburden

Bioburden testing establishes the number of microorganisms in a water sample, ensuring bacterial loads don't exceed mandated USP levels. The interesting thing about Bioburden testing is how would you even know you need to do it unless you waded through over 25 pages of USP<1231>! (Remember the little NOTE at the top of the monograph?)

Bioburden can be quantified by several methods: (1) pour plate, (2) spread plate and (3) membrane filtration. All these methods require adding a water sample to media and incubating the media for a few days and observing any growth.

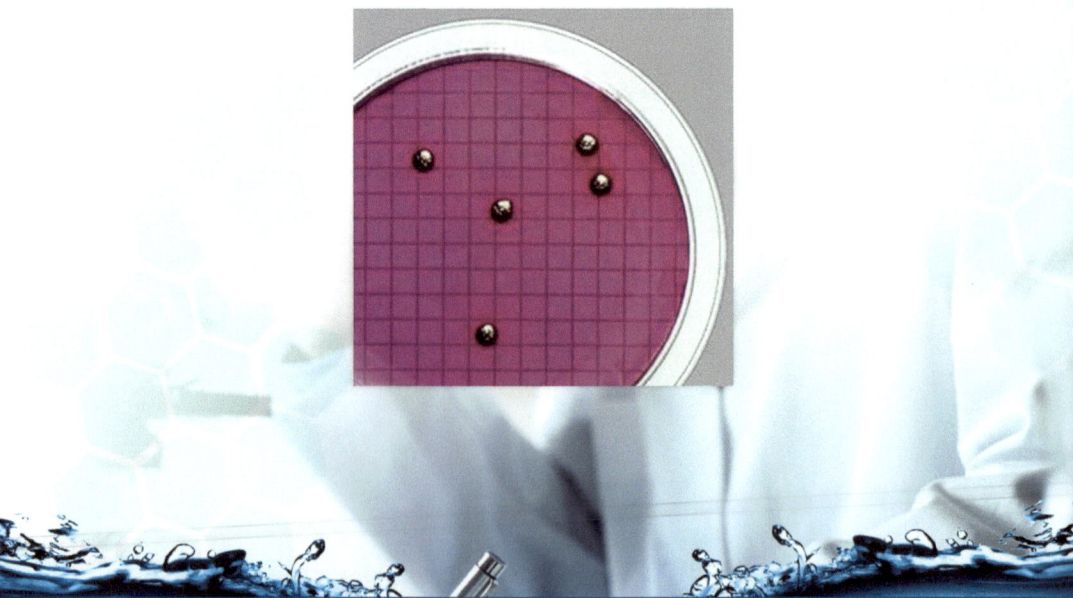

The #1 Source of Contamination

Most bacteria found during the testing of purified water is typically originated in the feed source water. This may be a result of worn-out filters on the system.

Let's Wrap It All Up… (and distill it all into one neat little test tube)

Well as far as your purified water is concerned there is apparently **"much ado about something"**!

We set out to give you an easy to understand explanation of

(1) what each of the purified water tests measure,

(2) how each test is performed, and

(3) some of the pitfalls in performing these tests.

If nothing else, you should now be in a better position to understand what's being done with the water that's collected in those sampling tubes, and be in a better position to understand what the test results are telling you. And, should your water fail anyone of the tests you now know some of the potential reasons.

But we want to take it one step further! We want to make your life easier, (aren't we nice!)

The Lesser of Two Evils

There are two predominate ways facilities such as yours have their purified water is tested -

It's either done in house where your own people perform the tests and generate the report,

Overstocking inventory with lab supplies

Running complicated test instruments with multiple people hours

Hours of of review cycles while de ciphering complicated data

or an independent lab comes to your facility, takes the samples, brings the samples back to

their lab, performs the tests, and generates the report.

Neither option is ideal. On the one hand your company is spending great deal of time,

money, and other resources to perform these tests in house, or you have to deal with the

disruption of having some group of outsiders "invade" your facility on a periodic and regular

basis. The decision becomes the lesser of two evils.

A Better Way to Do Water Testing!

We'd like to introduce you to our USP Purified Water Ulti-Pak:

All of Your Water Tests Done with One Single Vial!

We've Created a Solution Tailored for You

Ultimate Labs provides a one-of-a-kind service that meets all your USP, ISO, EPA and FDA requirements. Our integrated approach blends the benefits of both in-house and out-sourced testing.

You remain in control of the process and your workflow, while our specialized lab handles the complex work, specialized equipment, and compliance documentation.

We take care of the complex work — you remain in control.

Single test kit

5-minute sample collection

Real time results report

Our Work Here is Done

The quickest and easiest way to experience exactly what Ultimate Labs and the Ulti-Pak can do for you and your business is to jump on the Internet and type "http://waterultipak.com" into your web browser. Once there, you can view a brief description of our service. After picking up your jaw from your keyboard… watch our short video explaining how easy the process is. Then, we can set up a meeting (phone, web or in person) and walk you through the system and answer your questions about how we can seamlessly handle all your purified water testing **and save you a tremendous amount of time and money**.

BONUS MATERIAL #1

Here is an easy to read chart of different water types and their acceptance criteria. Enjoy!

	Purified Water		Highly Purified Water		Water For Injection	
	USP	EP	USP	EP	USP	EP
Process	Distillation, reverse osmosis and any other suitable process	Distillation, ion exchange, reverse osmosis and any other suitable process	N/A	Double-pass reverse osmosis coupled with other suitable techniques such as ultrafiltration and deionisation, for example	Distillation or reverse osmosis	Distillation
Conductivity	≤ 1.3 µS/cm @25°C	< 4.3 µS/cm @ 20°C	N/A	≤ 1.1 µS/cm @ 20°C	≤ 1.3 µS/cm @ 25°C	≤ 1.1 µS/cm @ 20°C
Bacteria	100 cfu/ml (suggested)	< 100 cfu/ml	N/A	<10 cfu/ 100 ml	< 10 cfu/ 100 ml (suggested)	< 10 cfu/ 100 ml
Endotoxin	N/A	< 0.25 IU/ml (only for bulk water for dialysis)	N/A	< 0.25 IU/ml	< 0.25 IU/ml	< 0.25 IU/ml
TOC	500 ppb	≤ 0.5 mg/l	N/A	≤ 0.5 mg/l	500 ppb	≤ 0.5 mg/l
pH	5-7	5-7	N/A	5-7	5-7	5-7
Nitrates	N/A	≤ 0.2 ppm	N/A	≤ 0.2 ppm	N/A	≤ 0.2 ppm
Heavy metals	N/A	≤ 0.1 ppm	N/A	≤ 0.1 ppm	N/A	≤ 0.1 ppm
Aluminium	N/A	≤ 10 ppb (if intended for use in the manufacture of dialysis solutions)	N/A	≤ 10 ppb (if intended for use in the manufacture of dialysis solutions)	N/A	≤ 10 ppb (if intended for use in the manufacture of dialysis solutions)

BONUS MATERIAL #2
Pass This Along as a Training Guide!

BONUS #3
Contact Us on How We Can Cut Your Testing Costs in Half!
Toll Free (858) 677-9297 and ask for Kim